M000236446

Managing Successful Science Fair Projects

Patricia Hachten Wee

J. WESTON
WALCH
PUBLISHER

Portland, Maine

User's Guide
to
Walch Reproducible Books

As part of our general effort to provide educational materials which are as practical and economical as possible, we have designated this publication a "reproducible book." The designation means that purchase of the book includes purchase of the right to limited reproduction of all pages on which this symbol appears:

Here is the basic Walch policy: We grant to individual purchasers of this book the right to make sufficient copies of reproducible pages for use by all students of a single teacher. This permission is limited to a single teacher, and does not apply to entire schools or school systems, so institutions purchasing the book should pass the permission on to a single teacher. Copying of the book or its parts for resale is prohibited.

Any questions regarding this policy or requests to purchase further reproduction rights should be addressed to:

Permissions Editor
J. Weston Walch, Publisher
321 Valley Street • P. O. Box 658
Portland, Maine 04104-0658

1   2   3   4   5   6   7   8   9   10

ISBN 0-8251-2804-8

For my parents,
Enid and Edward Hachten,
and my husband,
Robert

P.W.

Contents

Acknowledgments

*Thanks to the Media Center staff—
Deb, Candy, Peg, and Chris. Special
gratitude to Karen Leonhard for all her
assistance and to Janie Lloyd for her help.
And a thousand thank-yous to my husband
for his continuous support and great ideas!*

Teacher Introduction

This is actually a self-defense book for you, the teacher! I wrote it to help me keep up with the 100 science fair projects I can have in a year. When I began requiring projects of my science classes, I found myself facing the same problems you are undoubtedly encountering now:

- Which student has which project?

- Who is still determined to do a project on nuclear fusion?

- Who is so far behind that it's hopeless?

- Who is still waiting for you to get materials?

I looked at every science fair project book around and could find none that really helped me manage projects with the limited amount of time we teachers have. So, I wrote my own program. Taking the process step by step provided manageable segments to evaluate and an easy way to keep the students on track.

My students love the structure this book provides. They find that the due dates help them avoid procrastination and that the detail demanded in the experimental procedure is invaluable in producing a quality project. The exact format for writing the project report makes their work much easier. While many students worry about presenting their projects to the class prior to the science fair, all appreciate the "try-out" when the judging actually begins. Three of my students who were trained with this book have won some of the highest awards in their fields at the International Science and Engineering Fair.

The schedule of due dates I use has been included by the week. You can decide on the schedule that suits you best and fill in the dates before you copy the pages. If you have several classes, you may want to stagger the due dates to give you more flexibility in grading the worksheets. There is also a Student Progress Chart to help you keep track of each project. You can reproduce one chart for each class; you may want to post them to help students in tracking their own progress.

I have used two methods of distributing the pages to the students, both with equal success:

1. Copy the entire handbook at one time. This gives students the opportunity to work ahead.

2. Copy and distribute the introductory materials (Student Introduction, Science Fair Project Checklist, Science Fair Schedules), Tip Sheet 1, and Worksheet 1. In subsequent weeks, reproduce each numbered pair of Tip Sheets and Worksheets.

A final note: Students who want to work with vertebrate animals will need to use the *Guide for the Care and Use of Laboratory Animals* (NIH Publication 85-23), available from:

Office for Protection from Research Risks (OPRR)
National Institutes of Health
9000 Rockville Pike, Building 31, Room 5B63
Bethesda, MD 20892
(301) 496-7163

To the Student

The origin of science is the desire to know causes . . .
—William Hazlitt

A science fair project is an opportunity to experiment on your own, to investigate a piece of the real world, and to discover the thrill of real science. You have been preparing to play the game of science for many years in much the same way that athletes learn the rules and develop skills for a sport. Now is the time for you to get in the real game. To complete a science fair project, every student must follow the steps listed below:

1. Select a topic

2. Research the subject

3. Write the hypothesis

4. Plan the project

5. Experiment

6. Make observations

7. Analyze the results

8. Write the final project report and abstract

9. Prepare the exhibit

10. Prepare the presentation

As you proceed with your science fair project, you will be given tip sheets and worksheets to guide you through each of these steps. The checklist on pages 1 and 2 gives the date each item is due. Check off each step as you proceed. **Remember the due dates.** Place the graded worksheets in a notebook when they are returned to you; you will need to use them later.

Name _____ Date _____

Science Fair Project Checklist

(Check off each item when completed)

Week #   **Due Date**

SELECTING THE TOPIC

2   _____   ❒ Hand in Worksheet 1.

3   _____   ❒ Hand in Worksheet 2.

RESEARCHING THE SUBJECT

5   _____   ❒ Hand in Worksheet 3 and Research Notebook with at least six sources.

WRITING THE HYPOTHESIS

5   _____   ❒ Hand in Worksheet 4.

PLANNING THE PROJECT

8   _____   ❒ Hand in your checklist, Worksheet 5, and Worksheet 5a, if required.

DISTRICT SCIENCE FAIR PRELIMINARY RESEARCH PLAN AND CERTIFICATION

11   _____   ❒ **All students**—submit preliminary research plan to teacher.

11   _____   ❒ **Students with vertebrate animal/DNA/tissue/pathogen/ controlled substances/human study projects only**—submit the appropriate district science fair certification.

EXPERIMENTING (VERTEBRATE/HUMAN STUDY PROJECTS UPON APPROVAL BY DISTRICT SCIENCE FAIR)

11   _____   ❒ **Do not begin experimentation until your teacher has signed the "Approval to Proceed" line on the Teacher Approval Form.**

11   _____   ❒ Hand in Worksheet 6.

MAKING OBSERVATIONS

15   _____   ❒ Hand in Worksheet 7.

UNDERSTANDING THE FUNDAMENTALS

17   _____   ❒ Hand in Worksheet 8.

(continued)

Science Fair Project Checklist *(continued)*

Week #   **Due Date**

ANALYZING THE DATA

20   _____   ❑ Hand in Worksheet 9.

INTERPRETING THE RESULTS

22   _____   ❑ Hand in Worksheet 10.

SUBMIT ENTRY APPLICATIONS TO SCHOOL AND DISTRICT
SCIENCE FAIRS

23   _____   ❑ Submit completed entry applications to teacher for school and
district science fairs.

WRITING THE ROUGH DRAFT OF PROJECT REPORT—PART 1

23   _____   ❑ Hand in rough draft of title, introduction, hypothesis,
experimental procedure, and bibliography with Worksheet 11.

WRITING THE ROUGH DRAFT OF PROJECT REPORT—PART 2

24   _____   ❑ Hand in rough draft of observations, conclusion, and application
with Worksheet 11.

PLANNING THE EXHIBIT

25   _____   ❑ Hand in sketches of exhibit with Worksheet 12.

WRITING THE ROUGH DRAFT OF ABSTRACT

26   _____   ❑ Hand in rough draft of abstract with Worksheet 13.

PREPARING FOR THE ORAL PRESENTATION

27   _____   ❑ Hand in Worksheet 14.

27   _____   FINAL PROJECT REPORT AND ABSTRACT DUE

27   _____   PRESENT PROJECT TO CLASS

28   _____   SET UP PROJECT (SEE DETAILED SCHEDULE FOR SCHOOL
SCIENCE FAIR)

28   _____   SCHOOL SCIENCE FAIR (SEE DETAILED SCHEDULE FOR
SCHOOL SCIENCE FAIR)

30   _____   DISTRICT SCIENCE FAIR (SEE DETAILED SCHEDULE FOR
DISTRICT SCIENCE FAIR)

Name _____ Date _____

SCHEDULE FOR THE SCHOOL SCIENCE FAIR

Date _____

_____ P.M.   Set up projects in the _____.

Use the _____ entrance.

Date _____

_____ A.M.   Judging. All students must be present.

_____ P.M.   The science fair is open to the public.

_____ P.M.   Students remove their projects.

SCHEDULE FOR THE DISTRICT SCIENCE FAIR

Date _____

_____ P.M.   Set up projects in _____.

Date _____

_____ P.M.   Judging session. All students must be present.

_____ P.M.   Awards ceremony in _____.

_____ P.M.   District science fair is open to the public.

_____ P.M.   Students remove their projects.

Name _____ Date _____

Science Project Tip Sheet 1
SELECTING THE TOPIC

The universe is full of magical things patiently waiting for our wits to grow sharper.
—Eden Phillpots

Selecting a topic for your science fair project is probably the most difficult step. However, if you begin your search with a field of science in which you are already interested, this first step will be greatly simplified. The kind of project you need is one that produces measurable data; your project must deal with an effect that can be measured with numbers and units. The best projects pose a question and use scientific techniques to answer it.

Newspapers, science magazines, and science television programs offer many ideas for science fair projects. Your parents may also have some ideas for you. Topics may be found in the many books and magazines in your school library and in your science classroom. *The Reader's Guide to Periodic Literature* and the *SIRS Index* are excellent sources, as are the sources listed in Appendix D.

The following list shows the kind of ideas you can develop from a topic of interest. General topic: algae. Possible projects:

The Responses of Algae to Ultraviolet Light
The Effect of Metals on Transport Within Algae
The Effect of Water Pollution on Algae
The Sulfur Requirements of Algae Used for Human Food
Factors Influencing the Development of Colonies of Algae
A Study of Intracellular Algae in Hydras
How Uniform is the Thallus of *Ulva*, a Green Alga?

Keep in mind that your project will have to be limited in scope. You cannot do a project on all there is to know about algae as a food source. You would have to limit the range of your project, as shown in the list of projects on algae. We will use one of these projects, The Sulfur Requirements of One Type of Algae Used for Food, as a sample project as we go through the steps for a successful science fair.

Using vertebrate animals requires the completion of multipaged protocols that describe your entire experiment in great detail and involve at least one visit to a veterinarian. You will need to refer to the *Guide for the Care and Use of Laboratory Animals*. You will also need to complete numerous forms and certifications in order to use human subjects. The use of either vertebrate animals or humans will require the approval of the district science fair before you may begin your experimentation. There are many other organisms that would be just as interesting to work with, such as algae, ants, bacteria, beetles, crabs, crayfish, crickets, protists (*Daphnia*, or water fleas, cyclops, brine shrimp, hydra, etc.), fruit flies, houseflies, sowbugs, lichen, yeast, vinegar eels, slugs, snails, earthworms, planaria, viruses, and mealworms.

Science Project Worksheet 1

TOPIC SELECTION PARAGRAPH

Write one paragraph describing the idea you would like to pursue for your science fair project. Attach additional pages if necessary.

_____

_____

_____

_____

_____

_____

_____

_____

_____

_____

_____

_____

_____

_____

❏ The basic idea is workable, but you should

    ❏ limit the scope to:

_____

_____

_____

(continued)

Science Project Worksheet 1 *(continued)*
TOPIC SELECTION PARAGRAPH

❐ consider approaching it from a different angle, such as:

_____

_____

_____

❐ This idea sounds great (with these minor adjustments):

_____

_____

_____

❐ You need to find something entirely different and resubmit, because:

    ❐ this is too sophisticated for a high school project.

    ❐ this is an area not permitted by the district science fair.

    ❐ this is too elementary.

    ❐ this is not a data-gathering type of project.

Grade: _____

Name _____   Date _____

REFINING THE TOPIC

All science is concerned with the relationship of cause and effect.
—Lawrence J. Peters

Now that you have decided on a topic, you will have to develop a specific project. A research project deals with causes and effects. For example, in the sample project "The Sulfur Requirements of Algae Used for Human Food," the student would be growing a kind of algae that people use for food. The algae could be grown in solutions that have different amounts or kinds of sulfur. Some aspect of the sulfur requirement of the algae (the amount or kind) would be the *cause* and how well the algae grows would be the *effect*. You will need to compile a list of causes and effects for your topic with which you can work.

In a science fair project, the cause is called the *independent variable* and the effect is called the *dependent variable*. Usually there is only one independent variable in high school projects. In our example, the independent variable (the cause) could be either (a) the kind of sulfur (or from what chemical compounds the sulfur is taken), or (b) the amount of available sulfur.

In your research, you will change one factor in order to observe its effect on some other factor. The factor you change is the independent variable (the cause). The result obtained after manipulating the independent variable is the dependent variable (the effect). In the sample project, the following are some of the dependent variables that could be measured:

    a. growth rate of the algae

    b. protein content of the algae

    c. reproductive rate of the algae

    d. life span of the algae

Please note that both variables are factors that can be measured or recorded: the *kind* of sulfur, the *amount* of sulfur, the *rate* of growth, the *amount* of protein, the *rate* of reproduction, the *length* of life span.

Name _____ Date _____

Science Project Worksheet 2
REFINING YOUR TOPIC

Briefly describe the topic agreed upon in your Topic Selection Paragraph (Worksheet 1).

_____

_____

_____

_____

List four independent variables (causes) and four dependent variables (effects) in order of your preference. Begin with the variables that interest you most.

Independent Variables	**Dependent Variables**
a. _____	a. _____
b. _____	b. _____
c. _____	c. _____
d. _____	d. _____

Teacher Comments:

❏ Needs work. Resubmit by _____. Do more research to find more interesting variables.

❏ Well done! I agree with your preferences.

❏ Well done! However, the variables listed below will be more effective.

_____

_____

_____

_____

Grade: _____

Science Project Tip Sheet 3

RESEARCHING THE TOPIC

If I have been able to see farther than others,
it was because I stood on the shoulders of giants.
—Sir Isaac Newton

Once you have chosen your topic and received advisor approval, continue to read about your topic. An intensive review of all related literature is basic to all scientific research. To develop a quality project, you will need additional knowledge of the concepts and principles fundamental to your research, current information on the subject, and scientific support for your idea. You will need to document the accepted facts, concepts, and processes upon which the research is based. Your search of the scientific literature will enable you to find an area or particular effect that is not well known or completely researched.

Taking Notes

Organized notes make writing and drawing conclusions easier. When taking notes, record the reference information correctly and completely for use in your bibliography.

These are requirements for your Research Notebook:

1. Use a loose-leaf, three-ring notebook.

2. Write "Research Notes" on the cover.

3. Use ink to prevent smudging.

4. Begin your notebook with a title page. Use a tentative title at this point; it can be changed later.

5. Write the date on which you recorded the notes on each page.

6. Record the reference information for each new source consulted.

7. Write complete ideas. Be concise and clear. You may use abbreviations and your own shorthand. When recording data, draw a table or chart and fill it in.

8. Striking statements should be copied exactly and placed within quotation marks. This will enable you to use them later, if applicable, and give proper credit to the author.

Science Project Tip Sheet 3
RESEARCHING THE TOPIC *(continued)*

9. Do not hesitate to include articles that are not directly related to your ideas at this time; they may be pertinent later.

10. Always look for unique bits of information that would give you a "wow" focus for your project. (For example, a student whose topic was hydroponics discovered that the amount of boron required by plants is not known. Bingo!)

11. Set up each page of your Research Notebook according to the following format.

Sample Research Notebook Page

Date: 9/20/96

Article title: Growing Algae for Food

Author: R.G. Stanley and W.L. Butler

Title of Book: *Seeds: The Yearbook of Agriculture*

Publisher: Boston Publishing Inc., Boston, MA, 1995

Pages: 93–99

Useful Material: [Write all the notes in this section]

Significance to Project: [Make notes here to tag certain information and highlight specific points. For example:]

1. Confirms that sulfur is required for algae growth.
2. Gives basic explanation for raising algae.
3. Great chart! Possible use on display. [Attach sketch or photocopy.]

You will need to have completed research from six sources in your notebook by the due date listed on page 1. Hand in your Research Notebook with Worksheet 3 in front of the title page. More than six sources will be needed eventually, so do not stop researching.

(continued)

Name _____ Date _____

Science Project Worksheet 3
RESEARCH NOTEBOOK CHECKLIST

Use this checklist to be sure you have included everything. Place this worksheet in front of the title page of your research notebook.

❒ three-ring notebook
❒ "Research Notebook" written on cover
❒ title page
❒ written in ink
❒ proper format
❒ notes are dated
❒ titles recorded properly

❒ authors recorded properly
❒ publishers recorded properly
❒ page numbers are recorded
❒ notes are acceptable
❒ sources are relevant to the topic
❒ six or more sources

Teacher Comments:

_____

_____

_____

_____

_____

_____

❒ You have not completed this assignment satisfactorily. Resubmit by _____ .

❒ OK! Well done!

Grade: _____

Name _____ Date _____

WRITING THE HYPOTHESIS

The aim of science is to seek the simplest explanation of complex fact.
—Alfred North Whitehead

All scientific research must have a definite purpose or hypothesis. A science fair project is no exception! What question are you trying to answer? What experimental data are you seeking? You can only obtain limited answers in one science project. Be sure you limit your hypothesis to a goal you can attain. (For example, "A study of algae used for human food" may be too broad, while "A study of the sulfur requirements of one kind of algae" is much more feasible.)

A common mistake students make is choosing a purpose that involves more information, skill, and work than they are able to understand, attain, or complete in the time available. The choice of a project that requires materials or equipment that are unavailable in your school may mean arranging to work at a local university. This, though possible, entails more effort in scheduling, transportation, and actual work.

The hypothesis of your project should be stated in one sentence. Difficulty in accomplishing this indicates the need for further refinement or narrowing of the planned study. Several general forms can be used to state your hypothesis:

1. It is hypothesized that there is a direct relationship between _____ and _____.

2. The hypothesis of this research project is that _____ will cause a significant change in _____.

3. It is hypothesized that _____ will result in _____.

Following the forms above, the hypothesis of our sample project might be:

1. It is hypothesized that there is a direct relationship between the type of sulfur provided and the growth of algae.

2. The hypothesis of this research project is that certain forms of sulfur will cause a significant change in the growth rate of algae.

3. It is hypothesized that the kind and quantity of sulfur provided in the growth medium will result in different growth rates of algae.

(continued)

Science Project Tip Sheet 4

WRITING THE HYPOTHESIS *(continued)*

An additional benefit to a well-worded hypothesis is the ease with which a title can be created from the hypothesis. The title of your project should be sufficiently descriptive, but not too long. Try writing several titles before you come to any decision. A possible title for the sample project could be A Comparative Study of the Sulfur Requirements of Algae.

_____

_____

_____

_____

_____

_____

_____

_____

_____

_____

_____

_____

Science Project Worksheet 4

WRITING YOUR HYPOTHESIS

Write the hypothesis in the form that best reflects the goal of your project.

_____

_____

_____

_____

Tentative Project Title

_____

_____

Teacher Comments:

_____

_____

❏ You have not completed this assignment satisfactorily. Resubmit by _____ .

❏ OK! However, you will have a better project with the following hypothesis:

_____

_____

_____

❏ OK! Well done!

Grade: _____

Name _____ Date _____

PLANNING THE PROJECT

Science goes step by step.
—Sir Ernest Rutherford

A successful science fair project requires a good deal of careful planning. This planning will enable you to avoid some of the difficulties and pitfalls and to keep your project on track.

Your first step is to determine what equipment and supplies you will need. Check with your teacher to find out if the items are in stock. If they are not available, you must check science supply catalogs for the materials you need. It is your responsibility to find what you require. List the catalog name and date, the item number and name, and your name, and give the list to your teacher. You will be notified if there will be a cost to you before the materials are ordered. Most orders will be received within two weeks. You may need to call local universities for assistance with materials. Create a list of telephone numbers for the various departments of local colleges and universities.

A very important part of the plan for your project is deciding exactly what you will measure and exactly how you will measure it. You will need to discuss this aspect with your teacher, and you may need to check the catalogs again for the required measuring equipment. See the list of properties, units, and measuring instruments on pages 20 and 21.

In order to plan effectively, you will need to understand the role variables must play in your project. Explanations of each kind of variable follow.

Independent Variable

The *cause* of the change in your research is the *independent variable*. In your experimentation, you will vary one factor to observe what will happen. The thing you vary is the independent variable. In our sample project, the chemical that provides the source of sulfur is the independent variable.

Dependent Variable

You change one factor to observe what will happen to something else. The "something else" is the *dependent variable*, the *effect*. In our sample project, the dependent variable is the growth rate of the algae.

(continued)

Science Project Tip Sheet 5
PLANNING THE PROJECT *(continued)*

Controlled Variables

Only the independent variable is changed in a controlled experiment. All other factors must be kept constant and are called "controlled variables." In the sample project, the controlled variables are type and source of the algae, the temperature, the amount of light, the medium in which the algae grows (without the various forms of sulfur), etc.

Control Groups

Controlled experiments are designed to determine cause and effect. There must be a standard to which the experimental data can be compared. A control group is free of any change—free of the independent variable. The classic control group is the classroom of students who did not use Crest® toothpaste as compared to an experimental group that did. In the sample project, the control group is kept in the growth medium without any added sulfur, while the experimental groups are kept in the same growth medium with one sulfur-containing compound added. You must have a control group in order to compare the results of the experimental groups.

Your next job is to develop a sequence of steps, the experimental procedure, that will lead you through your project. Your procedure must include in detail everything you will do in your project. If you have chosen a project that involves vertebrate animals, be sure to refer to the requirements in the *Guide for the Care and Use of Laboratory Animals.* The experimental procedure is sufficiently detailed when another researcher can exactly duplicate your project by following your written plan.

This planning will allow you to set a schedule for yourself and give you a chance to anticipate some pitfalls and alternative methods. Remember to list the source of all materials and equipment. Keep in mind that all good scientific research is flexible enough to allow for changes and that some of the best learning comes through encountering and overcoming difficulties.

(continued)

Name _____ Date _____

PLANNING THE PROJECT *(continued)*

Staple your experimental procedure to Worksheet 5.

Note: Whether typing or writing your checklist, **double-space or skip a line.** Corrections and suggestions will be made in this space. Your final experimental plan will be more detailed after teacher evaluation and your continued research.

Here is the experimental procedure for our sample project.

Experimental Procedure Example

A Comparative Study of the Sulfur Requirements of Algae

- ❑ Oct. 5—arrange for experiment location (check if OK to take equipment home; ask at home if OK)
- ❑ Oct. 8—determine what kind of algae to use
- ❑ Oct. 15—order *Chlorella pyrenoidosa* from Carolina Biological Supply (stock #_____, page _____)—to arrive on _____
- ❑ Oct. 10—determine how to measure growth rate of the algae
- ❑ Oct. 12—procure supplies for nutrient growth medium:
  KNO_3 (from biology teacher)
  $MgCl_2$–$6H_2O$ (from biology teacher)
  KH_2PO_4 (from chemistry teacher)
  FeEDTA—order from Sigma Chemical Company (stock #_____, page _____)
  distilled water (chemistry teacher) save clean jugs for water
  H_3PO_4 (biology teacher)
  $MnCl_2$–$4H_2O$ (biology teacher)
  $ZnCl_2$ (biology teacher)
  $CuCl_2$–$2H_2O$ (chemistry teacher)
  Na_2SO_4 (chemistry teacher)
  Na_2SO_3 (chemistry teacher)
  $Na_2S_2O_3$ (order, Carolina Biological Supply, stock #___, page ___)
  Na_2S (order, Carolina Biological Supply, stock #___, page ___)
  sterilizer (biology teacher)
  60 test tubes (biology teacher)
  test tube marker (biology teacher)
  cotton for stoppers (buy at drug store)

(continued)

Science Project Tip Sheet 5

PLANNING THE PROJECT *(continued)*

water bath at 25° C:

    thermostatically controlled heater (biology teacher)

    small aquarium (biology teacher)

glass tubing (chemistry teacher) about 5 cm/test tube (total 3 m)

Tygon tubing (chemistry teacher) about 5 cm/test tube (total 3 m)

CO_2 generator (biology teacher)

air pump, small—(order from Wards, stock #_____, page ____)

2 fluorescent bulbs and fixtures (biology teacher)

light meter (photography teacher)

photoelectric colorimeter w/600mµ filter—(order from Carolina Biological Supply, stock #_____, page ___)

☐ Oct. 15—compare costs of sulfur sources

☐ Oct. 20—review and write up all procedures for growing *Chlorella* and operating colorimeter

☐ Oct. 24—set up tubing assemblies

☐ Oct. 30—run tests to achieve correct gas rates for CO_2 and air

☐ Nov. 1—determine from library research how fast *Chlorella* is supposed to grow

☐ Nov. 6—set up water bath and check temperature maintenance

☐ Nov. 7—check positioning of test tubes in water bath: will they fit?

☐ Nov. 8—determine distance of lights to achieve 600 foot-candles on test tubes

☐ Nov. 10—test colorimeter with solutions of green food coloring in test tubes

☐ Nov. 10—arrange for club time or after school with teacher to mix growth solutions and to sterilize tubes and solutions

☐ Nov. 15—mix solutions:

control:	*micronutrient solution:	experimental media:
1.21 g KNO_3	2.86 g H_3BO_3	A: add 0.02% Na_2SO_4
2.03 g $MgCl_2–6H_2O$	1.81 g $MnCl_2–6H_2O$	B: add 0.02% Na_2SO_3
1.23 g KH_2PO_4	0.105 g $ZnCl_2$	C: add 0.02% $Na_2S_2O_3$
2 ml FeEDTA	0.055 g $CuCl_2–2H_2O$	D: add 0.02% Na_2S
1 ml micronutrient solution*	distilled H_2O to a 1-liter volume	

☐ Nov 16—set up water bath, CO_2 generator, air pump, lights, etc. Do trial set-up to be sure all is ready.

☐ Nov 17—fill test tubes with media and sterilize

☐ Nov 18—set up charts for growth data

(continued)

Science Project Tip Sheet 5

PLANNING THE PROJECT *(continued)*

❐ Nov 19—decide on method to measure initial amount of algae to put in each test tube

❐ Nov 20—when *Chlorella* arrives, set up control and experimental groups:
control: 12 test tubes
A: 12 test tubes   C: 12 test tubes
B: 12 test tubes   D: 12 test tubes

❐ Nov 20—make initial colorimeter measurement and repeat hourly for 6 hours

❐ Nov 21—take colorimeter readings at 8-hour intervals (**Note:** this schedule may be modified depending upon results of research.)

❐ Nov 22—research (1) if possible to chemically determine if *Chlorella* converts sulfide into sulfite or sulfate before using the sulfur; (2) if possible to determine protein content of *Chlorella*. Call universities for info.

❐ Nov 25—repeat entire experiment with greater concentration of sulfur (0.04%)

❐ Dec 8—repeat entire experiment with lesser concentration of sulfur (0.01%)

❐ Dec 18—repeat entire experiment with another concentration of sulfur—depending on results of previous experiments.

❐ Dec 28—continue experimenting along the paths revealed by the data

❐ Jan 28—work on graphs

❐ Jan 30—work on statistics

❐ Feb 2—finalize statistical work

❐ Feb 3—set aside time to work on graphs and statistical analyses

❐ Feb 9—work on rough draft of project report

❐ Feb 10—set aside library time for further research

❐ Feb 11—work on exhibit (punch out letters, decide on colors and props)

❐ Feb 15—work on exhibit and rough draft of abstract

❐ Feb 19—work on rewrite of project report

❐ Feb 19—rewrite project report

❐ Feb 20—photocopy graphs and charts and enlarge for backboards

❐ Feb 23—rewrite abstract

❐ Feb 25—finish exhibit

❐ Mar 1—set up project at Science Fair

❐ Mar 2—Science Fair

Name _____ Date _____

Properties, Units, and Measuring Instruments

In order to vary the independent variable, to measure the dependent variable, and to control all other variables, the researcher must use a wide variety of laboratory instruments. Many of the properties you will need to measure are in the following list. The metric unit in which the property is measured, the abbreviation for the unit, and some of the instruments used to measure the property are also given.

Property	Measurement Unit	Abbrev.	Instruments Used
Acidity/Alkalinity	pH		pH paper, pH meter
Angle	degrees	(°)	protractor, sextant, transit
Area	$meter^2$ $centimeter^2$ $millimeter^2$	(m^2) (cm^2) (mm^2)	meter stick and formula for regular objects, planimeter for irregular objects or indirect measurement for irregular objects
Density	$kilogram/meter^3$ $grams/centimeter^3$	(kg/m^3) (g/cm^3)	balance and meter stick, pycnometer, hydrometer
Electrical Current	ampere	(amp)	ammeter
Electrical Potential	volt	(V)	voltmeter
Electrical Resistance	ohm	(Ω)	ohmmeter, Wheatstone bridge
Force	newton	(N)	spring scale
Growth (special): Optical density Size of colony	nanometer number per square millimeter	(nm) $(\#/mm^2)$	photoelectric colorimeter marked grid or overlay
Heat	joule	(J)	calorimeter
Humidity	percent	(%)	hygrometer
Length	meter centimeter millimeter micrometer Angstrom	(m) (cm) (mm) (μm) (A)	meter stick, tape measure, micrometer, vernier caliper
Light intensity	candle lumen	(can) (lum)	photometer, light meter, photoelectric cell
Mass	kilogram gram milligram microgram	(kg) (g) (mg) (μg)	spring balance, lever-arm balance, electronic balance

(continued)

Name _____   Date _____

Properties, Units, and Measuring Instruments
(continued)

Property	Measurement Unit	Abbrev.	Instruments Used
Pressure	pascal pounds/square inch	(N/m^2) (lb/in^2)	barometer, manometer, mechanical pressure gauge
Sound intensity	decibel	(db)	audiometer, sound level meter
Temperature	degrees centigrade or Celsius	(^{o}C)	thermometer, thermocouple, thermistor, pyrometer
Time	seconds	(s)	stopwatch
Velocity	meter/second	(m/s)	speedometer, anemometer, stopwatch and meter stick
Volume	cubic meter cubic centimeter cubic millimeter	(m^3) (cm^3) (mm^3)	graduated cylinder, pipette, burette, volumeter, manometer
Weight	newton	(N)	spring scale

Name _____   Date _____

Science Project Worksheet 5
PLANNING YOUR PROJECT

Equipment

1. The dependent variable is _____

The units it will be measured in are _____

It will be measured with (the instruments to be used and the number of significant figures each will measure):

Instrument	# of significant figures
_____	_____
_____	_____
_____	_____

2. The independent variable is _____

The units it will be measured in are _____

It will be measured with (the instruments to be used and the number of significant figures each will measure):

Instrument	# of significant figures
_____	_____
_____	_____
_____	_____

3. List all the controlled variables and how you will keep them constant.

_____

_____

_____

_____

(continued)

                   Managing Successful Science Fair Projects

Science Project Worksheet 5

PLANNING YOUR PROJECT *(continued)*

4. Describe your control group.

_____

_____

_____

_____

5. List the organisms you plan to use and their scientific names (scientific names are in Latin and are, therefore, <u>underlined</u> or *italicized*).

_____

_____

_____

_____

6. If you are using vertebrate animals, complete Worksheet 5a—Vertebrate Animal Requirements.

Safety

List **all** chemicals you plan to use.

_____

_____

_____

_____

(continued)

Name _____   Date _____

PLANNING YOUR PROJECT *(continued)*

Grade Sheet for Worksheet 5 and 5a

Teacher Comments:

❏ dates are realistic   ❏ scientific name of organism

❏ control acceptable   ❏ sufficient depth to project

❏ sufficient detail for control   ❏ sufficient detail in checklist

❏ independent variable   ❏ all chemicals are listed

❏ dependent variable   ❏ Worksheet 5a attached if necessary

❏ controlled variables   ❏ scientific names listed

❏ control group   ❏ food

❏ specified factors to be measured   ❏ cage or confinement

❏ units for each measurement   ❏ bedding, if applicable

❏ specified technique for measurements

❏ Needs work. You have not completed this assignment satisfactorily.
    Resubmit by _____ .

❏ OK! Well done!

Grade: _____

Name _____ Date _____

VERTEBRATE ANIMAL REQUIREMENTS

List all the requirements for the vertebrates you are planning to use. Refer to the *Guide for the Care and Use of Laboratory Animals* (NIH Publication 85-23), available from the Office for Protection from Research Risks (OPRR), National Institutes of Health, 9000 Rockville Pike, Building 31, Room 5B63, Bethesda, MD 20892.

Vertebrate animal (common name): _____

Scientific name (**Note:** the first word is the genus name and is capitalized; the second word is the species name and is not capitalized. The entire scientific name is *italicized* or underlined.):

_____

Food: _____

Food specified in *Guide for the Care and Use of Laboratory Animals*:

_____

Cage or confinement area size: _____

Cage or confinement area size specified in *Guide for the Care and Use of Laboratory Animals*:

_____

Bedding material: _____

Bedding material specified in *Guide for the Care and Use of Laboratory Animals*:

_____

Specifications of vertebrates to be used:

Age: _____ Sex: _____ Number: _____

Genetic background: _____

Animal supplier: _____

Cost: _____

Name _____ Date _____

Science Project Teacher Approval Form

Project Title _____

After you have implemented the suggestions/changes made on Worksheet 5 and after inspecting your research apparatus, materials, and methods, my recommendation is:

❏ You MAY NOT proceed with your experimentation until

_____

_____

_____

_____

_____

_____

❏ You MAY proceed with your experimentation as described above.

_____

_____

_____

_____

Approval to Proceed _____ Date_____

Name _____   Date _____

Science Project Tip Sheet 6
EXPERIMENTING

Take nobody's word for it; see for yourself.
—The motto of the Royal Society

Working on your science fair project at home or in the laboratory is an interesting and rewarding experience. During your research, you will be working with materials and equipment that can cause injury if not handled properly. Accidents do not just happen—they are caused by carelessness, haste, and by ignoring safety rules. Before beginning your lab work, read the following safety rules, learn them, and adhere to them.

General Rules

1. Experimenting is serious work, not horse-play.

2. Perform only those experiments previously approved by your teacher.

3. Keep your work area clean and uncluttered.

4. Wear safety goggles when working with flame or chemicals.

5. Know the location of the fire extinguisher, eyewash station, and safety shower before you begin work.

6. Have all apparatus checked out by your teacher before use.

7. Use the correct tool for each job.

8. Keep combustible materials away from heat and flame.

9. Never put your face near an open container of chemicals.

10. Clean up all spills immediately.

11. Report all accidents immediately to your teacher, if working at school, or to your parents, if working at home.

12. Experiments involving fumes or vapors must be conducted under a fume hood.

13. Do not view lasers, the sun, ultraviolet lamps, or other light sources directly.

(continued)

27   *Managing Successful Science Fair Projects*

Name _____ Date _____

Science Project Tip Sheet 6

EXPERIMENTING *(continued)*

Biological Precautions

1. Use only nonpathogenic bacteria and fungi.

2. Seal all petri dishes with tape and never open them.

3. Wear rubber gloves when working with bacteria.

4. Kill all cultures of bacteria and fungi before disposing of them. This can be done in an autoclave or sterilizer.

5. Make sure that you follow all rules for working with vertebrate animals.

Chemical Precautions

1. Use only chemicals you have obtained from your teacher.

2. Read and double-check all labels on chemical bottles. Take only as much as you need.

3. Avoid touching chemicals.

4. Do not use unlabeled chemicals.

5. Do not return unused chemicals to stock bottles.

6. Never taste chemicals.

7. Use of known or suspected carcinogenic chemicals is strictly regulated. Ask your teacher.

8. Pour acid into water—NEVER pour water into acid.

9. Never mix chemicals without prior clearance from your teacher.

(continued)

Name _____ Date _____

Science Project Tip Sheet 6

EXPERIMENTING *(continued)*

Recording Data

For your research data to have significance, you will have to be aware of the measuring limitations of the instrument used. However, you will be able to produce meaningful research results by consulting your teacher about accuracy, significant figures, and scientific notation.

Note: You will need to keep an experimental notebook for your data. Another loose-leaf, three-ring notebook with Experimental Notebook written on the cover is required for your exhibit. Enter all your data and observations in ink. Date all entries. You should also include costs, supplies, problems encountered, and how you solved those problems.

The experimental procedure for your project can be in list or paragraph form. Use the past tense, passive voice. (For example, instead of "I mixed the solution," say "the solution was mixed.") The scientific point of view is impersonal. Avoid use of the first person (*I, me, mine, etc.*). Refrain from using contractions. No separate listing of materials is to be made. Materials will be incorporated into the procedural steps.

The test subjects (plants, animals, protists) must be identified by common name and by scientific name. The scientific name is written in italics or underlined. The first word of the scientific name (the genus) is capitalized, but not the second (the species). Specify the source of the test subjects. (For example, Corn seeds, *Zea mays*. Obtained from Parks Brothers Seed Company, taken from 1995 harvest.)

Specify the exact type and brand of all test equipment. If the equipment was borrowed from the school, identify it as such, but do not indicate the name of the school.

Clearly indicate the independent and dependent variables, the controlled variables, and the control group. Use the term "experimental group(s)" for your test group(s).

Another science student of your caliber should be able to exactly duplicate your research from the procedure you have written.

Name _____ Date _____

EXPERIMENTAL PROCEDURE

Write your experimental procedure in the form specified on page 29 so that it can be used for your project report. Use the checklist below to keep on track while you are writing the procedure and again while you are proofreading. Staple your experimental procedure to this worksheet.

- ❐ Proper form (list or paragraphs)
- ❐ Past tense, passive voice is used ("Seeds were planted...").
- ❐ First-person pronouns (*I, me, my*) are not used.
- ❐ Contractions are not used.
- ❐ Test subjects (plants, animals, protists) are named by:
    - ❐ common name
    - ❐ scientific name (genus and species are underlined or italicized)
- ❐ Source of test subjects is specified.
- ❐ Control group is specified.
- ❐ Experimental group(s) are specified.
- ❐ Independent variable(s) are described.
- ❐ Dependent variable(s) are described.
- ❐ Test equipment specified by:
    - ❐ brand
    - ❐ model
    - ❐ source (if borrowed from school, do not use school's name)
- ❐ Controlled conditions are described.
- ❐ Another student could duplicate the experiment from this procedure (have another science student read it to be sure).

(continued)

Science Project Worksheet 6

EXPERIMENTAL PROCEDURE *(continued)*

Teacher Comments:

❒ Proper form

❒ Past tense, passive voice

❒ Test subjects named

❒ Common name of test subjects

❒ Scientific name of test subjects

❒ Control group specified

❒ Experimental groups specified

❒ Independent variable(s) described

❒ Dependent variable(s) described

❒ Test equipment specified

❒ Controlled conditions are sufficiently described.

❒ Another student could duplicate the experiment from procedure given.

❒ Resubmit by _____ .

❒ OK! Well done!

Grade: _____

Name _____ Date _____

Science Project Tip Sheet 7
MAKING OBSERVATIONS

In the field of observations, chance only favors
those minds which have been prepared.
—Louis Pasteur

Now that you are conducting your experiments, you will need to develop methods to record the data generated. The following table is one simple format that can be used in your experimental notebook. After making adjustments to your data tables, you will use copies in your final project report and on your display.

Table 1

Growth of *Chlorella* in Media of Various Sulfur-containing Compounds
(measured in nm of optical density)

Test tube # Control	Optical Density (nm)	Test tube # Exp. A (Na_2SO_4)	Optical Density (nm)
1	10.5	1	12.8
2	9.9	2	12.4
3	10.1	3	12.5
4	10.3	4	12.4
5	10.0	5	12.7
Mean	10.2	Mean	12.6

The title of the table names the dependent variable first and then the independent variable. It clearly identifies the relationship between the two variables (here, "Growth of *Chlorella* in media of various sulfur-containing compounds"). The title of a table shows the unit of measurement utilized (nm, nanometers). Each column must be labeled with the appropriate information. Tables are consecutively numbered.

As a general rule, your data must be graphed. A graph requires a title that names the dependent variable first, then the independent variable. The two axes must be labeled with the factor represented and the units of measurement. Generally the independent variable is on the *X*-axis and the dependent variable is on the *Y*-axis. An appropriate scale must be used on each axis. There are many different types of graphs, each of which is appropriate for certain kinds of data or to emphasize certain relationships. Graphs are consecutively lettered.

There are computer programs that will turn your data into graphs. Try these programs as soon as you have some data. This will give you enough time to try different formats.

Name _____ Date _____

Science Project Worksheet 7

RESULTS AND OBSERVATIONS

Project Title _____

    Fill in the following data tables and sketch graphs with titles, units, etc. If you do not yet have actual data to include, you may leave the columns of measurements blank for now. Use the checklist to keep on track. Attach additional pages if necessary.

Table 1

(Title) _____

_____ **(measured in _____)**

Mean		Mean	

Table 2

(Title) _____

_____ **(measured in _____)**

Mean		Mean	

(continued)

Name _____   Date _____

Science Project Worksheet 7
RESULTS AND OBSERVATIONS *(continued)*

Graph A   **Graph B**

Checklist for tables

- ❑ Title names the dependent variable first.
- ❑ Title names the independent variable second.
- ❑ Title identifies the relationship between the dependent variable and the independent variable.
- ❑ Title includes the unit of measurement.
- ❑ Each table is numbered consecutively.
- ❑ Each column is labeled appropriately.

(continued)

Science Project Worksheet 7

RESULTS AND OBSERVATIONS *(continued)*

Checklist for graphs

❑ Title names the dependent variable first.

❑ Title names the independent variable second.

❑ The dependent variable is labeled on the *Y*-axis.

❑ The independent variable is labeled on the *X*-axis.

❑ *Y*-axis is labeled with the unit of measurement.

❑ *X*-axis is labeled with the unit of measurement.

❑ *Y*-axis has an appropriate scale.

❑ *X*-axis has an appropriate scale.

Teacher Comments:

_____

_____

_____

_____

❑ Resubmit by _____ .

   ❑ data tables are not correct

   ❑ graphs are not correct

❑ OK! Well done!

Grade _____

Science Project Tip Sheet 8

UNDERSTANDING THE FUNDAMENTALS

Science is organized knowledge.

—Herbert Spencer

In order to draw valid conclusions from your experiment, you must completely understand all the scientific concepts involved in your project.

The student with the sample project "A Comparative Study of the Sulfur Requirements of *Chlorella*" would need to have complete knowledge of the following terms and concepts:

terms	concepts
algae	the metabolism of *Chlorella*
Chlorella	the importance of *Chlorella* to humans
mμ	the forms of sulfur used by *Chlorella*
nm	what is different about the various forms of sulfur
foot-candle	the role of sulfur compounds in the metabolism of *Chlorella*
sulfur	equivalent C and F temperatures
sulfide	differences between fluorescent and incandescent lights
sulfite	how a light meter works
sulfate	how a photoelectric colorimeter works
protein	
FeEDTA (and other unusual chemicals)	

Name _____   Date _____

Science Project Worksheet 8

BACKGROUND INFORMATION

Write from three paragraphs to three pages of background information. Explain **every** concept, scientific principle, etc., that you have used or referred to in your project. These explanations must be complete and detailed. Attach additional pages if necessary.

_____

_____

_____

_____

_____

_____

_____

_____

_____

_____

_____

_____

_____

_____

_____

_____

_____

(continued)

Science Project Worksheet 8

BACKGROUND INFORMATION *(continued)*

Teacher Comments:

_____

_____

_____

_____

❑ Resubmit by _____ .

   ❑ not enough depth to the explanations of the following concepts:

_____

_____

_____

   ❑ not all concepts/scientific principles have been explained. Include the following:

_____

_____

_____

❑ OK! Well done!

Grade _____

Science Project Tip Sheet 9
ANALYZING THE DATA

Numerical precision is the very soul of science.
—Sir D'Arcey Wentworth Thompson

To determine what your data really mean, it is essential to analyze the results statistically. How different must the experimental data be from the control data in order to mean something? Statistical analysis of your data will answer this and other questions. While statistics may initially seem difficult, most high school students can learn to use them quite effectively.

Discuss data analysis and statistics with your advisor. Many computer programs will perform statistical analyses. You may need to borrow statistics books from the mathematics department or the library in order to understand what the calculations mean and how they are done. Some useful books are:

Brase, Charles Henry, and Corrinne Pellillo Brase. *Understandable Statistics.* Lexington, MA: D.C. Heath and Company, 1987.

Sincich, Terry. *Statistics by Example.* San Francisco: Dellen Publishing Co., 1990.

Sokal, Robert R., and F. James Rohlf. *Biometry: The Principles and Practice of Statistics in Biological Research.* San Francisco: W. H. Freeman, 1981.

You will want to incorporate such statistical analyses as chi-square, "t" test, and standard deviation into your project. Do not procrastinate on starting this part of your project; it can be time-consuming.

After calculating the statistics, the null hypothesis, and the significance of your data, you must decide what all this means. Have your results occurred by chance or are your findings of significance? In the sample project, you would say "according to the data generated by this research, *Chlorella* provided with sulfur from Na_2SO_4 grew at a significantly greater rate ($p<0.001$) than the control group or other experimental groups."

Your analysis may indicate the need for further experimentation. The statistical results below would require that more research be done with algae grown in Na_2SO_4 and Na_2S in order to provide conclusive data:

control vs. Na_2SO_4 $p< 0.2$

control vs. Na_2S $p< 0.2$

         Managing Successful Science Fair Projects

Name _____ Date _____

STATISTICS

Set up tables for your project statistics as directed by your teacher. You must have all the work for $\bar{x}$, s^2, s, "t" test and degrees of freedom (if applicable), and chi-square (if applicable). Plot a distribution curve for the standard deviation. Attach this worksheet to your papers.

1. Are your data discrete or continuous? _____

2. Should your calculation be "t" test or chi-square? _____

3. What is the number of degrees of freedom? _____

4. What is your null hypothesis? _____

   _____

5. Do you reject or accept the null hypothesis? _____

   At what level of confidence? _____

6. Are your findings significant, highly significant, or not significant? _____

7. State your findings.
   According to the data generated by this research, _____

   _____

   _____

   _____

   _____

8. Do you need to do further experimentation? _____

   If yes, on which aspect? _____

(continued)

Science Project Worksheet 9
STATISTICS *(continued)*

Teacher Comments:

_____

_____

_____

_____

❑ Resubmit by _____ .

   ❑ incorrect determination of type of data

   ❑ incorrect choice of test

   ❑ stats are incorrect

   ❑ stats are missing

   ❑ incorrect number of degrees of freedom

   ❑ null hypothesis is incorrect or inadequate

   ❑ incorrect level of confidence

   ❑ incorrect classification of significance

   ❑ incorrect statement of findings

   ❑ distribution curve incorrect

❑ OK! Well done!

Grade _____

Science Project Tip Sheet 10
INTERPRETING THE RESULTS

Science is simply common sense at its best—that is, rigidly accurate in observation and merciless to fallacy in logic.
                                                —Thomas H. Huxley

The purpose of scientific research is to determine if there is a cause-and-effect relationship between the independent and dependent variables, and whether that relationship is based on something other than chance.

To draw a logical conclusion from experimental research, it is necessary to analyze the data. The analysis of experimental results requires three processes: complete understanding of the concepts upon which the research is based, statistical analyses of the data, and logical interpretation of the meaning of the analyses.

Worksheet 8 provided an opportunity to check your knowledge of the background information necessary to analyze your results. Statistical analysis is presented in Worksheet 9.

The next step in analysis is determining the reason(s) for the results. Why did what happened happen? From our example, why did Na_2SO_4 cause significantly greater growth? What is it about Na_2SO_4 that promotes the growth of these algae? Why did Na_2S, $Na_2S_2O_3$, and Na_2SO_3 not do the same? How are these compounds chemically different? You will probably need to do more library research to find the answers. Do not forget to add the new sources to your bibliography.

Name _____ Date _____

Science Project Worksheet 10
INTERPRETING YOUR RESULTS

State the results and then determine the reason(s) for the results. Why did what happened happen?

According to the data generated in this research project, _____

_____

_____

_____

_____ resulted in

_____

_____ .

These results may have been due to _____

_____

_____

_____

Teacher Comments:

_____

_____

_____

❏ Resubmit by _____.

    ❏ explanation of the results missing or inadequate

    ❏ explanation of reasons for results missing or inadequate

  ❏ OK! Well done!

Grade _____

Science Project Tip Sheet 11

WRITING THE PROJECT REPORT

Science is nothing but developed perceptions, integrated intent, common sense rounded out and minutely articulated.

—George Santayana

The purpose of your science fair project report, and of any scientific paper, is to persuade the reader that the conclusions you have drawn are correct. This goal can be accomplished if you write clearly and concisely.

Your project report must be typed. Margins should be $1\frac{1}{2}$ inches on the left and 1 inch on the right, top, and bottom. When typing your report, double-space each section. You may use footnotes if appropriate.

Your project report must include:

1. Title Page
1. Introduction
1. Hypothesis
1. Experimental Procedure
1. Results and Analysis

1. Conclusions
1. Application
1. Bibliography
1. Acknowledgments

Each section must begin on a separate page. The content of each of these sections and specifications for writing them are as follows.

Title Page   The title should be centered on the page. No identification of the student or the school should appear.

Introduction   The introduction is placed on the page immediately following the title page. There is no number attached to the title "Introduction." The introduction to your paper should be no longer than one page.

The objective of the introduction is to capture the interest of a reader knowledgeable about the field of study, but not the specific problem. A good introduction begins with a discussion of the source of your topic. Explain your choice, particularly if an article or news item influenced you or if you had an interesting personal reason. Discuss the significance of your topic. Briefly cite the research of other scientists that specifically relates to your project. Conclude the introduction with a clear, concise statement of the reason your research was done, and include the significant implications of your study and exactly what hypothesis was being tested.

(continued)

Science Project Tip Sheet 11

WRITING THE PROJECT REPORT *(continued)*

The introduction will also offer a ready-made, brief synopsis of your project (perfect for answering the typical opening question of a judge, "Tell me about your project").

Use the introduction to convince the reader that your research was of sufficient importance to have undertaken it!

Hypothesis   The hypothesis of the project report is written in the body of the report and prefaced by Roman numeral I. It must be a complete sentence.

Experimental Procedure   The experimental procedure for the project is also written in the body of the report and is prefaced by Roman numeral II. The procedure can be in list or paragraph form. Use the past tense, passive voice (e.g., "The solution was mixed," not "I mixed the solution."). The scientific point of view is impersonal. Avoid the use of first person pronouns *(I, me, mine)*. Refrain from using contractions. No separate listing of materials is to be made. Materials will be incorporated into the procedural steps.

The test subject must be identified by common name and by scientific name. The scientific name is written in italics or underlined. The first word of the scientific name (the genus) is capitalized, but not the second (the species). Specify the source of the test subjects (e.g., corn seeds, *Zea mays.* Obtained from Parks Brothers Seed Company, taken from 1995 harvest).

Specify the exact type and brand of all test equipment. If the equipment was borrowed from the school, identify it as such, but do not indicate the name of the school.

Clearly indicate the independent and dependent variables, the controlled variables, and the control group. Use the term "experimental group(s)" for your test group(s).

Another science student of your caliber should be able to exactly duplicate your research from the procedure you have written.

Results and Analysis   The results and analysis section is written in the body of the report and prefaced by Roman numeral III. Use the past tense, passive voice "Measurements of the stem diameter were taken . . ." not "I measured the diameter of the stems . . .").

This section is a presentation of the data gathered during the experimentation and the statistical analysis. As such, it includes the work done in Worksheets 7, 9, and 10.

(continued)

Science Project Tip Sheet 11

WRITING THE PROJECT REPORT *(continued)*

You will discuss the data and refer to data tables and graphs. Each table and graph should be on a separate page following the text portion of your results and analysis. Each table and graph must be clearly and completely titled, labeled, and scaled, and have the units marked. Each table should be consecutively numbered and each graph consecutively lettered.

You may discuss your analyses separately after the discussion of all your results or with each section of results. All the analyses should have separate sheets of calculations. Your analyses should be explained and discussed in the body of the results and analysis section. Refer to the sheets of calculations, each of which needs to be labeled completely and titled. Be certain to explain the acceptance or rejection of the null hypothesis for readers not knowledgeable in the field of statistics.

Conclusion   This section is written in the body of the report and is prefaced by Roman numeral IV. The conclusion explains how you interpreted your results and statistical and graphical analyses. The conclusion should be first stated in one sentence that parallels your hypothesis in content and grammatical structure. Several paragraphs can then be used to further explain your conclusions. Reference should be made to the similarities and differences between your findings and interpretations and the work of others.

Application   The implications of your research are explained in this section. It will not be numbered with a Roman numeral. Judges often like to see that your research has some specific value to life in your county or state (to its industry, farming, water quality, etc.).

Bibliography   The bibliography appears on a separate page and will not be numbered with a Roman numeral. Your teacher, the librarian, or an English teacher can provide a copy of the correct bibliographic form. If you have used most of a book, you do not need to list page numbers.

Remember that the bibliography is alphabetized according to the first word of each entry. Sources are not numbered.

Acknowledgments   This section appears on a separate page (without a Roman numeral) or on a card displayed with your project. This section gives the names of all the people who helped you with your project. Include parents, teachers, and other professionals who assisted you. This section should be added or displayed only after the judging is completed.

Science Project Worksheet 11

PROJECT REPORT CHECKLIST

The rough draft of your project report is due in two sections. Part 1 includes the Title Page, Introduction, Hypothesis, Procedure, and Bibliography. Part 2 includes the Results and Analysis, Conclusion, and Application. These rough drafts may be handwritten in pencil, but must be double-spaced to allow for corrections. As you proceed, use the checklist below; attach it to your rough draft with a paper clip.

Title Page

- ❐ Title contains the dependent and independent variables. If you have chosen an attention-getting title, you may omit this requirement.
- ❐ Title is centered.

Introduction

- ❐ Introduction begins on a new page.
- ❐ Introduction is not preceded by a Roman numeral.
- ❐ Describes source of research idea.
- ❐ Describes other relevant research on your topic.
- ❐ Answers the question, "Why was your research done?"

Hypothesis

- ❐ Hypothesis is on a separate page.
- ❐ Hypothesis is preceded by Roman numeral I.
- ❐ Hypothesis is a complete sentence.

Experimental Procedure

- ❐ Procedure begins on a new page.
- ❐ Procedure is preceded by Roman numeral II.
- ❐ Identifies test subjects, with both common and scientific names.
- ❐ Describes control and experimental groups.
- ❐ Dependent and independent variables are described.
- ❐ Test equipment is specified.
- ❐ Describes methods used to control variables.
- ❐ Gives concise description of experimental method.
- ❐ Another student could completely duplicate this experiment from this description.

(continued)

Science Project Worksheet 11
PROJECT REPORT CHECKLIST *(continued)*

Results and Analysis

❏ Results and Analysis begins on a new page.

❏ Is preceded by Roman numeral III.

❏ Written in past tense, passive voice.

❏ All data are given in table and/or graphic form on separate pages.

❏ Graphs and tables are correctly titled, labeled, numbered, or lettered.

❏ All data are discussed.

❏ All calculations for statistical analyses are shown.

❏ Statistical analyses are discussed, including null hypothesis, etc.

❏ Reasons are offered to explain the results.

Conclusion

❏ Conclusion begins on a new page.

❏ Conclusion is preceded by Roman numeral IV.

❏ Conclusion is stated in one sentence that is parallel with Hypothesis.

❏ Results are interpreted.

❏ Comparison(s) made to work of other scientists, if applicable.

Application

❏ Application begins on a new page.

❏ Application is not preceded by a Roman numeral.

❏ Connection to your county or state or general application is made.

Bibliography

❏ Bibliography begins on a new page.

❏ Bibliography is not preceded by a Roman numeral.

❏ Bibliography is alphabetized by first word.

❏ Entries are in the correct form.

Acknowledgments (need not be included with your rough draft)

❏ Begins on a new page or on a separate card.

❏ Is not preceded by a Roman numeral.

❏ Will be added or displayed *after* the judging.

(continued)

Science Project Worksheet 11

PROJECT REPORT CHECKLIST *(continued)*

Part 1 (Title Page, Introduction, Hypothesis, Procedure, Bibliography)

Teacher Comments:

_____

_____

_____

_____

❏ Resubmit by _____. See checklist for errors.

❏ OK! Well done. Proceed with your final copy.

Grade: _____

Part 2 (Results and Analysis, Conclusion, Application)

Teacher Comments:

_____

_____

_____

_____

❏ Resubmit by _____. See checklist for errors.

❏ OK! Well done. Proceed with your final copy.

Grade: _____

Science Project Tip Sheet 12

PLANNING THE EXHIBIT

A picture is worth a thousand words.

Now that your research and paper are completed, it is time to share your findings with the judges and the public. You have three opportunities to report your research: your project report and abstract, your oral presentation to the judges, and your exhibit.

Your exhibit must be self-explanatory, as it will represent you and your project. Part of the judging process takes place without you, so your display must speak for you. The key to an effective display is planning.

You have seen many kinds of displays in the photographs and actual exhibits shown by your teacher or advisor. Creativity and neatness are essential. Use photographs if possible. Try to make your display striking and unforgettable! Do not display anything within 6 inches of the bottom of your backboards, as this area is too low to be used effectively. A typical exhibit design is sketched below.

Your exhibit must have the title, hypothesis, observation, graphs/data, conclusion, and statistics displayed on the backboards. You may include the application and photographs. If you have constructed any of the equipment used, make this fact clear on the exhibit. Photographs of your construction of such equipment and the parts used are essential. Refer to the entry in this section under Photographs on page 52.

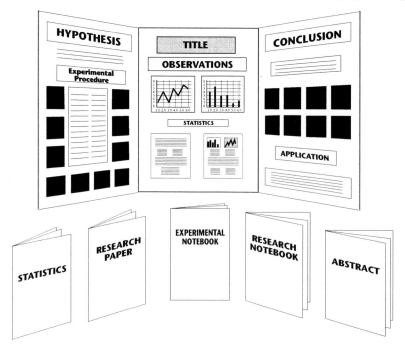

Science Project Tip Sheet 12

PLANNING THE EXHIBIT *(continued)*

Backboard

A rigid portable backboard is required. To construct your backboard, the district science fair regulations must be followed (for example, height—9 feet floor to top [most tables are 3 feet high]; width—4 feet; depth—30 inches). The backboard must be self-standing and self-supporting. It can be constructed from many kinds of materials (pegboard, foam board, plywood, etc.) and should be three-sided with hinges.

You may want to cover the backboard. Coverings can include burlap, paint, construction paper, poster board, and wallpaper.

Posters

To prevent time-consuming mistakes, make your display on individual small posters that can be attached to your backboard. Many students use construction paper and poster board. If you wish, you can purchase colored poster board at local stationery or discount stores. Be creative and attentive to color schemes. Combinations of colors are often effective.

Lettering

Avoid hand-lettering unless you are an artist. You can borrow or buy stencils in many sizes. It is most efficient to draw the letters on construction paper or poster board and cut them out to paste on your posters. Coloring in stenciled letters with markers can be messy and defeat your goal of neatness. Your school or an elementary school may have a letter-punch which you may be able to borrow. Pasting of cutout letters is best accomplished with rubber cement or glue stick. You may choose to purchase commercial peel-off/stick-on letters, but they are quite expensive. Computer graphics programs can also produce large headings and lettering for your display. Avoid using one long banner for your title unless you can mount it on a rigid support.

For writing the text of your introduction, purpose, experimental procedure, results and analysis, conclusion, and application on your posters, make a second copy of your computer printout or typed project report and have it enlarged on the photocopier. It may also be possible to have the computer print a copy of the text of your report in a larger type.

Headings and text can be mounted on colored construction paper and then attached to your posters. Always try to construct your display in small removable segments so that portions can be changed easily. You may need to make some changes in your display before entering the district science fair. If each segment is separate and easily detached, you will be able to make such changes quickly and with little effort.

(continued)

Science Project Tip Sheet 12

PLANNING THE EXHIBIT *(continued)*

Graphs, data tables, and statistics

Your graphs, tables, and statistics will need to be mounted on your display. Most of your data display will be generated with the computer programs. Hard copies of your graphs and tables can be photocopied, or the originals can be used in your project report. You can make a second copy of your graphs and tables on the photocopier and enlarge them at the same time. These should be mounted as explained in the preceding paragraph.

Photographs and videorecordings

"One picture is worth a thousand words." Try to use photographs of your procedure, set-up, and results. Pictures are a "must" if you have constructed a significant part of your project. Mount photographs on construction paper or poster board with circles of adhesive tape, for easy removal. If your photographs show a sequence of events, place the earliest one at the upper left. For other photographs, place the most attention-getting picture in the center with others grouped around it.

It may be desirable to use enlargements of some of your photographs. Decide on this as soon as possible, since it may take weeks to have enlargements made. Don't use photographs that in any way identify you or your school. Refer to photography books for assistance in taking and displaying your photographs.

It may also be possible to use a videorecording of parts of your experimentation. This type of presentation will require a VCR, a small television set, and an electrical outlet. Consult your teacher about the practicality of this method.

Specimens and props

There will be quite a large table area in front of your backboard. Use this to your best advantage. A colorful table covering is effective. Arrange important pieces of equipment, specimens, and models carefully to make your display attractive and attention-getting. Be certain to include your Experimental Notebook, Research Notebook, and Abstract. Follow the district science fair rules for items that cannot be displayed.

You should also pack a small box or bag with supplies to make last-minute repairs to your backboard when you set up your project. Include masking tape, clear tape, duct tape, scissors, nails, hammer, screwdriver, pliers, stapler, thumbtacks, glue, pens, eraser, extra hinges, extra construction paper, and poster board that matches the poster board in your project.

Name _____ Date _____

Science Project Worksheet 12

EXHIBIT CHECKLIST

1. Sketch your ideas for your exhibit on pieces of plain paper and attach them to this worksheet. Use one sheet for each board of your backboard, one sheet to show what you will display on the table, and as many others as are needed for graphs, etc. All can be roughly sketched. Graphs should show titles, scales, and labels, but the actual graph can be approximated.

2. Indicate your choice of colors.

   _____

3. What material will cover your backboards?

   _____

4. Indicate what technique you will use for lettering.

   _____

5. Check the space required for your title and the technique you plan to use (letter-punch, stencil, computer graphics). What is the height and width of the area designated for your title?

   _____

   What size lettering will you use for your title?

   _____

   How much space will this size lettering take to spell out your title?

   _____

Teacher Comments:

❏ Resubmit by _____. It needs further work on the following:

   _____

   _____

❏ Terrific ideas! Good luck in completing your exhibit.

Grade _____

Name _____ Date _____

Science Project Tip Sheet 13

WRITING THE ABSTRACT

*Language is only the instrument of science,
and words are but the signs of ideas.*
—Samuel Johnson

The abstract is a 250-word summary of your entire project. It must include your hypothesis, a brief explanation of your experimental procedure, a brief overview of your observations, your conclusion, and applications. You should not include graphs or tables. Your conclusion need not be a specific statement that parallels your hypothesis, but can be combined with applications.

Your abstract must identify the dependent and independent variables and the research technique used. It may not include any information not included in your research report.

As with all scientific writing, your abstract must be written in past tense, passive voice. You will write "A water bath was prepared . . . ," not "I made a water bath . . . ."

Use your abstract to convince the reader that your research was of sufficient interest to have undertaken it. This may be the only part of your project the judges read. Their first impression of your display and your abstract determines your success!

Reducing your entire project to 250 words is a task that will require a good deal of time. A sample abstract based on the example project used throughout this book follows.

A Comparative Study of the Sulfur Requirements of *Chlorella*

The hypothesis of this project was that the kind and quantity of sulfur provided to *Chlorella pyrenoidosa*, a green alga used as a human food supplement, will directly affect the growth rate. *Chlorella* requires sulfur, but not all sulfur-containing compounds can be utilized. This research project was designed to provide some data on the types of sulfur compounds effective in widespread algae farming.

Chlorella was grown under controlled conditions of 600 foot-candles of light, 2.5 cm^3/sec CO_2 and 50 cm^3/sec air at 2 lb/in^2, 25°C in sterilized test tubes of nutrient medium. Experimental groups were grown in the nutrient medium with the addition of one of the following sulfur-containing compounds at 0.02%: Na_2SO_4; Na_2SO_3; $Na_2S_2O_3$; Na_2S.

(continued)

       Managing Successful Science Fair Projects

Science Project Tip Sheet 13

WRITING THE ABSTRACT *(continued)*

Growth in the culture tubes was measured using a Ward photoelectric colorimeter with a 600 mµ filter. Growth, in optical density, was plotted against time, in hours. Growth rates of the various groups were compared using the "t" test with the following results:

control group vs. Na_2SO_4 = $p < 0.001$

control group vs. $Na_2S_2O_3$ = $p < 0.01$

control group vs. Na_2SO_3 = $p < 0.2$

control group vs. Na_2S = $p < 0.2$

From these results, it may be determined that the growth rate of *Chlorella* is significantly affected by differences in sulfur sources and that Na_2SO_4 provides the optimal requirements for growth.

Name _____ Date _____

Science Project Worksheet 13
ABSTRACT CHECKLIST

As you write your abstract, use the checklist below. Be sure to double-space. After each paragraph, write the total number of words in that paragraph in the margin. Staple this sheet to your rough draft.

- ❏ Double-spaced or written on alternate lines

- ❏ Begins with a statement of hypothesis

- ❏ Gives some background

- ❏ Written in past tense, passive voice

- ❏ Includes a brief summary of research plan

- ❏ Contains a brief statement of results

- ❏ Includes a summary of statistical analyses and significance

- ❏ Gives a brief conclusion and/or application

- ❏ Number of words in each paragraph written in the margin

- ❏ Does not exceed 250 words

- ❏ Total number of words is _____

Teacher Comments:

_____

_____

_____

_____

- ❏ Your abstract needs more work. Resubmit by _____.

- ❏ Excellent! You may make the final copy of your abstract with the corrections given.

Grade _____

Name _____ Date _____

Science Project Tip Sheet 14
PREPARING FOR THE ORAL PRESENTATION

Science is analytical, descriptive, informative.
—Eric Gill

The oral presentation of your project is your best opportunity to show what you know and what you have accomplished. Your best preparation will be practicing at home in front of your family or in front of a mirror. Making a video or audiotape of your presentation will help you find areas that need improvement.

Dress appropriately for the judging session and for the evening public exhibit. Your teacher will tell you what type of clothing is customary. You may want to bring a snack, but keep it unobtrusive. **Do not chew gum**. These guidelines must also be followed for the district judging session and awards ceremony.

During the judging presentation, you will have your exhibit to use as a visual device and to help you remember facts and figures. Use it! Remain at your project at all times. Do not wander. You will have an opportunity to see the other students' projects during the evening public exhibit.

Normally, judges visit you individually and may stay from one minute to thirty minutes. You may have only one judge, or as many as twenty! When a judge arrives at your exhibit, stand up and smile. If the judge does not immediately ask you a question (typically, "Tell me about your project"), begin with your well-rehearsed opening statements. Some suggestions for opening statements will be given later. Some judges like to appraise your exhibit and even read your abstract before talking with you. If a judge seems to want to follow that path, allow it graciously.

Be courteous and polite at all times. If the judge offers advice on improving your project, listen carefully and thank him or her. You will want to keep a pen and tablet handy to jot down comments and advice given by the judges. These bits of information may help you improve your project for the next year or suggest alternate methods of study.

Do not be upset if the judge asks you a question that you cannot answer. Simply say that you do not know. Do not try to fake it. However, if you feel that the question involves some aspect that you did not intend to research or some factor that you were unable to measure, explain that politely.

(continued)

Name _____ Date _____

Science Project Tip Sheet 14
PREPARING FOR THE ORAL PRESENTATION *(continued)*

Your opening statements are very important and are easy to master. Tell the judge what is exciting and unique about your project. If you had a special or personal reason for selecting this project, explain it. Convince the listener that this is a great project and that you really accomplished something. Show enthusiasm! If you master three to five minutes of opening statements, you will put yourself at ease for the remainder of the judging session.

Here are some tips for your oral presentation to the judges:

- Know your project completely.

- Do not memorize your presentation.

- Dress neatly and professionally.

- Relax.

- Stand up when a judge arrives.

- Be enthusiastic.

- Speak slowly and clearly.

- Be polite.

- Make eye contact.

- Smile.

- Keep it simple.

- Be concise.

- Take a deep breath if you get confused.

- If you don't know, say so.

- Be proud of your project. Remember that not many students can accomplish what you have!

58   *Managing Successful Science Fair Projects*

Name _____ Date _____

ORAL PRESENTATION CHECKLIST

Use the checklist below to judge your oral presentation. Practice in front of a mirror or make the presentation to friends or family; make a video or audio recording, or present your project to your class or advisor.

- ❐ I did not chew gum.
- ❐ I made use of the display.
- ❐ I had a three-to-five-minute opening statement.
- ❐ I used correct grammar.
- ❐ I smiled (at least sometimes).
- ❐ I had a tablet and pen handy.
- ❐ I did not fake answers.
- ❐ I took a deep breath if I got lost or confused.
- ❐ I spoke slowly and clearly.
- ❐ I was enthusiastic.
    I practiced:
    - ❐ with a video camera.
    - ❐ with an audio recorder.
    - ❐ in front of a mirror.
    - ❐ in front of friends or family.
    - ❐ in front of a class.

Teacher Comments:

_____

_____

_____

_____

❐ Your oral presentation needs more work. Resubmit by _____.

❐ Excellent! Good luck with your oral presentation to the judges. Keep practicing.

Grade _____

For Further Reading

American Mathematical Society. *A Manual for Authors of Mathematical Papers*. 7th ed. Providence, RI: American Mathematical Society, 1980.

Andersen, Arden B. *Science in Agriculture*. Kansas City, MO: Acres USA, 1992.

Barlow, Roger J. *Statistics: A Guide to the Use of Statistical Methods in the Physical Sciences*. New York: Wiley, 1989.

Behringer, Marjorie Perrin. *Techniques and Materials in Biology*. Malabar, FL: Robert E. Krieger Publishing Company, 1986.

Beller, Joel. *Experiments with Plants*. New York: Prentice Hall Press, 1985.

Bishop, Cynthia, Katherine Ertle and Karen Zeleznik, eds. *Science Fair Project Index, 1985-1989*. Metuchen, NJ: Scarecrow Press, 1992.

Brase, Charles Henry, and Corrinne Pellillo Brase. *Understandable Statistics*. Lexington, MA: D.C. Heath and Company, 1987.

Dodds, John H., and Lorin W. Roberts. *Experiments in Plant Tissue Culture*. 2nd ed. New York: Cambridge University Press, 1986.

Freund, John E. *Statistics: A First Course*. Englewood Cliffs: Prentice Hall, 1991.

Hanlin, Robert T., and Miguel Ulloa. *Atlas of Introductory Mycology*. Winston-Salem, NC: Hunter Textbooks, Inc., 1979.

Harris, C. Leon, ed. *Tested Studies for Laboratory Teaching*. Dubuque, IA: Kendall/Hunt Publishing Company, 1984. 4 vols.

Kreiger, Melanie Jacobs. *How to Excel in Science Competitions*. New York: Franklin Watts, 1991.

Kyte, Lydiane. *Plants from Test Tubes*. Portland, OR: Timber Press, 1987.

Lehman, Richard S. *Statistics and Research Design in the Behavioral Sciences*. Belmont, CA: Wadsworth Publishing Co., 1991.

Loiry, William. *Winning with Science*. Sarasota, FL: Loiry Publishing, 1990.

Micklos, David A., and Greg A. Freyer. *DNA Science*. [Burlington, NC]: Carolina Biological Supply Company and Cold Spring Harbor Laboratory Press, 1990.

(continued)

FOR FURTHER READING *(continued)*

Millspaugh, Ben. *Aviation and Space Projects*. Blue Ridge Summit, PA: TAB Books, 1992.

Morholt, Evelyn, and Paul F. Brandwein. *A Sourcebook for Biological Sciences*. New York: Harcourt Brace Jovanovich, Publishers, 1986.

Moore, David S., and George P. McCabe. *Introduction to the Practice of Statistics*. New York: W.H. Freeman and Co., 1989.

Newton, David E. *Consumer Chemistry Projects for Young Scientists*. New York: Watts, 1991.

Newton, David E. *Making and Using Scientific Equipment*. New York: Watts, 1993.

O'Neil, Karen. *Health and Medicine Projects for Young Scientists*. New York: Watts, 1993.

Post, Frederick J. *A Laboratory Manual for Food Microbiology and Biotechnology*. Belmont, CA: Star Publishing Co., 1988.

Sanders, Donald H. *Statistics: A Fresh Approach*. New York: McGraw-Hill, 1990.

Schuler, Mary A., and Raymond E. Zielinski. *Methods in Plant Molecular Biology*. New York: Academic Press Inc., 1989.

Sincich, Terry. *Statistics by Example*. San Francisco: Dellen Publishing Co., 1990.

Smoothey, Marilyn, and Ted Evans. *Statistics*. New York: Marshall Cavendish, 1993.

Sokal, Robert R., and F. James Rohlf. *Biometry: The Principles and Practice of Statistics in Biological Research*. San Francisco: W. H. Freeman, 1981.

Tant, Carl. *Science Fair Spelled W-I-N*. Angleton, TX: Biotech Publishing, 1992.

Tobias, Sheila. *Succeed with Math*. New York: College Entrance Examination Board, 1987.

Van Deman, Barry A., and E. McDonald. *Nuts and Bolts: A Matter of Fact Guide to Science Fair Projects*. Harwood Heights, IL: The Science Man Press, 1980.

Wall, Francis J. *Statistical Data Analysis Handbook,* New York: McGraw-Hill, 1986.

Watson, James D., John Tooze, and David T. Kurtz. *Recombinant DNA: A Short Course*. New York: W. H. Freeman and Company, 1983.

Yoshioka, Ruby, ed. *Thousands of Science Projects*. Washington, DC: Science Service, 1987.

Appendix A
SCIENCE FAIR PROJECT TOPIC IDEAS

1. What is the relationship between weeds and soil nutrient deficiencies?

2. How dependent are magnetic bacteria on specific iron sources?

3. A study of social hierarchy among fish

4. Can crop yields be increased by the use of methanol?

5. Yeasts and bacteria as natural fungicides

6. Agricultural by-products to clean wastewater

7. Seed/plant oils as natural fungicides/pesticides

8. A study of the antimicrobial effects of seeds

9. Alum and other chemicals as solutions to the serious phosphorus overload caused by use of poultry manure as fertilizer

10. Comparison of alternative farming systems on water quality

11. Computer modeling of animal population fluctuations

12. Cooling with sound

13. Utilization of the resazurin reduction method for determination of biological activity in sludge

14. Removal of atmospheric formaldehyde by plants

15. Effects of increased UV light on phytoplankton

16. Effects of aquatic toxins on freshwater invertebrates

17. *In vivo* lethality of brine shrimp as a measure of bioactivity

18. A comparison of reaction time of athletes vs. non-athletes

19. Can a natural fungus inhibitor be extracted from strawberries and raspberries?

20. Use of a gravity table to separate low- and high-protein wheat

(continued)

SCIENCE FAIR PROJECT TOPIC IDEAS *(continued)*

21. Induction of tomato leaf lesions and potassium deficiency by excessive ammonium nutrition

22. A study of lead in water of homes built before 1988

23. A measurement of lead leaching from pottery and leaded crystal

24. Zebra mussel control

25. Acid pollution and low-calcium bird egg shells

26. A study of the natural antibiotic in parsley seeds

27. A study of yeast and bacteria to protect fresh fruit and cut flowers

28. Can garlic produce better memory in rats?

29. Coral reef health as it relates to excessive salt

30. The effects of gibberellic acid on the growth of crop plants

31. Growth characteristics of bacteria mutants in various media

32. Oxygen diffusion rates of different soils

33. Alkalinity in limestone soils

34. Exploration of germination requirements of native plants

35. A study of autofluorescence of plant cells when damaged

36. A study of the use of bacteria to guard against dry rot fungus

37. Forestalling citrus fruit shrinkage by external coatings

38. Can reduced tillage increase soil carbon and boost crop production on sandy soils?

39. Using modeling to study streambed erosion

40. Can herbicide use be decreased by changes in planting density?

(continued)

SCIENCE FAIR PROJECT TOPIC IDEAS *(continued)*

41. A study of water and electrolyte exchange in animal tissues

42. Designing a more efficient wind-powered generator

43. A comparison of various measures to prevent soil erosion

44. The effects of plant extracts on tumor growth in crown gall disease

45. Can electromagnetic radiation affect the maze performance of rats?

46. Lectins: to study cell surfaces

47. Bioluminescence of ostracods and bacteria

48. Stress-testing of building materials

49. What factors affect the performance of oil-eating bacteria?

50. Can various odors affect student performance on tests?

Appendix B
EQUIPMENT AND MATERIALS SUPPLIERS

Aldrich Chemical Company, Inc.
1001 West Saint Paul Avenue
Milwaukee, WI 53233
1-800-558-9160
(chemicals)

Carolina Biological Supply Co.
2700 York Rd.
Burlington, NC 27215
1-800-334-5551
(scientific equipment, biological
supplies, live specimens)

Connecticut Valley Biological Supply
  Co., Inc.
82 Valley Rd., P.O. Box 326
Southampton, MA 01073
1-800-628-7748
(scientific equipment, earth science
supplies)

Difco Laboratories
P.O. Box 331058
Detroit, MI 48232-7058
1-800-521-0851
(microbiology products)

Edmund Scientific Co.
101 E. Gloucester Pike
Barrington, NJ 08007
1-609-573-6250
(scientific equipment)

ENSYS
P.O. Box 14063
Research Triangle Park, NC 27709
1-800-242-7472
(test kits for hydrocarbons, PCBs,
pesticides, metals)

Fisher Scientific
4901 W. LeMoyne
Chicago, IL 60651
1-800-955-1177
(scientific equipment, physics and
chemistry supplies)

Frey Scientific Co.
905 Hickory Lane
Mansfield, OH 44905
1-800-225-3739
(scientific equipment)

Grau-Hall Scientific Corporation
6501 Elvas Ave.
Sacramento, CA 95819
1-800-331-4728
(scientific equipment, molecular
biology supplies)

Nasco
901 Janesville Ave.
Fort Atkinson, WI 53538-0901
1-800-558-9595
(scientific equipment, biological
supplies)

PASCO Scientific
1876 Sabre St.
Hayward, CA 94545
1-800-722-8700
(scientific equipment, physics
supplies)

Promega Corporation
2800 Woods Hollow Road
Madison, WI 53711-5399
1-800-356-9526
(biological research products,
supplies for work with enzymes,
proteins, molecular biology, genetic
analysis)

(continued)

EQUIPMENT AND MATERIALS SUPPLIERS *(continued)*

Sargent-Welch
7400 N. Linder Avenue
Skokie, IL 60077
1-800-727-4368
(scientific equipment)

Science Kit & Boreal Laboratories
777 East Park Drive
Tonawanda, NY 14150-6784
1-800-828-7777
(scientific equipment, live specimens,
biology supplies)

Showboard, Inc.
3725 W. Grace St., 305
Tampa, FL 33607
1-800-323-9189
(science fair supplies, display boards,
titles)

Sigma Chemical Company
P.O. Box 14508
St. Louis, MO 63178-9916
1-800-325-3010
(chemicals)

Synthephytes Division
Plant Something Different, Inc.
22318 South County Road, 48
P.O. Box 1032
Angleton, TX 77516-1032
1-800-659-3078
(biotechnology lab kits, books)

Ward's Natural Science Establishment, Inc.
P.O. Box 92912
Rochester, NY 14692-9012
1-800-962-2660
(scientific equipment, biological
supplies, live specimens)

Appendix C
MAJOR SCIENCE COMPETITIONS

Duracell Scholarship Competition

Grades 9–12
Students must build a working device powered by batteries.
Deadline: Late January
Scholarships awarded
Sponsored by Duracell USA and NSTA
Address for information: Duracell Scholarship Competition
     National Science Teachers Association
     1742 Connecticut Avenue, NW
     Washington, DC 20009

International Science and Engineering Fair

Grades 9–12
Students exhibit completed independent research projects in science, engineering, or mathematics.
Deadline: Entrance form deadlines vary by region.
     Regional competitions are generally in February or March.
     International competition is in May.
Scholarships, cash, trip to Nobel Prize Ceremonies, and other awards
Sponsored by Science Service
Address for information: Science Service
     1719 N Street NW
     Washington, DC 20036

Junior Academy of Science

High school students
Students present completed scientific or mathematical research papers.
Deadline: Entrance form deadlines vary by region.
     Regional competitions are generally held in February.
     National competition is generally held in February or March.
Scholarships, trip to the National Association of Academies of Science Annual Meeting, and other awards
Sponsored by National Association of Academies of Science
Address for information: Gloria Takahashi
     American Junior Academy of Science
     Southern California Academy of Science
     900 Exposition Boulevard
     Los Angeles, CA 90007
     (213)744-3384 or (818)333-2173

Appendix C
MAJOR SCIENCE COMPETITIONS *(continued)*

Junior Science and Humanities Symposium

Grades 9–12
Students submit completed science or humanities independent research papers.
Deadline: Regional deadlines vary by area
       National deadline: May
Scholarships and trip to the London International Science Youth Fortnight
Sponsored by the Academy of Applied Science and the U.S. Army Department of Research
Address for information: Academy of Applied Science
              98 Washington Street
              Concord, NH 03301

Space Science Student Involvement Program

Grades 7–12
Student submit various types of project proposals or essays. The format changes from year to year.
Deadline: usually March
Certificates and trips awarded
Sponsored by NASA and NSTA
Address for information: Space Science Student Involvement Program
              National Science Teachers Association
              1742 Connecticut Ave., NW
              Washington, DC 20009

State Science Talent Searches

High school seniors
Students submit independent science research papers.
Deadline: December
Scholarships awarded
Sponsored by Science Service
Address for information: Science Service
              1719 N Street NW
              Washington, DC 20036

(continued)

Appendix C

MAJOR SCIENCE COMPETITIONS *(continued)*

The Thomas A. Edison/Max McGraw Scholarship Program

Grades 9–12
Students submit completed independent research projects in science or engineering or a proposal for such a project.
Deadline: December 15
Sponsored by the Max McGraw Foundation and the National Science Education Leadership Association
Scholarships awarded
Address for information: National Science Education Leadership
Association
c/o Copernicus Hall, Room 227
Central Connecticut State University
1615 Stanley Street
New Britain, CT 06050

Westinghouse Science Talent Search Competition

High school seniors
Students submit a completed independent research project.
Deadline: December
Sponsored by the Westinghouse Electric Company and Science Service
Address for information: Science Service
1719 N Street NW
Washington, DC 20036

Appendix D
REFERENCE SOURCES FOR
SCIENCE FAIR PROJECTS

Resources for Scientific Information

Gordon, Arnold J., and Richard A. Ford. *The Chemist's Companion: A Handbook of Practical Data, Techniques, and Reference.* New York: Wiley, 1972.

Hanson, A. A., ed. *Practical Handbook of Agricultural Science.* Boca Raton, FL: CRC Press, 1990.

Legge, Allan H. and Sagar V. Krupa. *Air Pollutants and Their Effects on the Terrestrial Ecosystem.* New York: Wiley, 1986.

Lynch, Charles T., ed. *Practical Handbook of Materials Science.* Boca Raton, FL: CRC Press, 1989.

Mason, Robert J., and Mark T. Mattson. *Atlas of the United States Environmental Issues.* New York: Macmillan, 1990.

O'Leary, William M., ed. *Practical Handbook of Microbiology.* Boca Raton, FL: CRC Press, 1989.

Databases of Scientific Information

Biology Digest

> Organizes, summarizes, and indexes worldwide scientific literature in the life sciences. Published on the 15th of each month except June, July, and August by Plexus Publishing, Inc., 143 Old Marlton Pike, Medford, NJ 08055

Magazine Articles Summaries (M.A.S.)

> M.A.S. provides access to articles in over 400 general magazines from 1984 to the present. Updated monthly by: EBSCO Publishing, P.O. Box 2250, Peabody, MA 01960-9765 (508)535-8545

(continued)

Appendix D
REFERENCE SOURCES FOR
SCIENCE FAIR PROJECTS *(continued)*

NewsBank

> Collection of articles from over 500 U.S. newspapers reprinted on microfiche and indexed on CD-ROM. Science-related categories include the environment, land development, and science and technology. Updated index and articles published monthly by NewsBank Inc., 58 Pine Street, New Canaan, CT 06840-5426 1-800-243-7694

Online Computer Library Center (OCLC)

> Public library database. Updated monthly from 6565 Frantz Road, Dublin, OH 43017

Scientific American Cumulative Index 1978-1988

> Index to 126 issues of *Scientific American* magazine from July 1978 through December 1988. Published by Scientific American, Inc., 415 Madison Avenue, New York, NY 10017 1-800-333-1199

SIRS (Social Issues Resources Series)

> Loose-leaf volumes covering 32 social issues and two critical issues which include some science-related topics. Articles are reprinted from newspapers, magazines, government documents and journals. Twenty articles published annually by Social Issues Resources Series, Inc., P.O. Box 2348, Boca Raton, FL 33427-2348 (407) 994-0079

SIRS Science

> Reprinted articles in five major categories: Earth, Life, Physical, Medical and Applied, which reflect the latest developments and issues in the field. Seventy articles published annually by Social Issues Resources Series, Inc., P.O. Box 2348, Boca Raton, FL 33427-2348 (407) 994-0079

(continued)

REFERENCE SOURCES FOR
SCIENCE FAIR PROJECTS *(continued)*

Periodicals and Journals

Ag Research
Published by the U.S. Department of
    Agriculture
Superintendent of Documents
P.O. Box 371954
Pittsburgh, PA 15250-7954

Chem Matters
American Chemical Society
1155 16th St., NW
Washington, DC 20036

ChemEcology
Chemical Manufacturers Association
2501 M Street, NW
Washington, DC 20037

Chem 13 News
Department of Chemistry
University of Waterloo
Waterloo ON N2L 3G1
Canada

Futures
Published by the Michigan State
    University Agricultural Experiment
    Station
310 Agriculture Hall
Michigan State University
East Lansing, MI 48824
(517)336-1555

Journal of Chemical Education
Department of Chemistry
University of Wisconsin-Madison
1101 University Avenue
Madison, WI 53706-1396

Penn State Agriculture
Agricultural Information Service
The Pennsylvania State University
106 Agricultural Administration
University Park, PA 16802

The Physics Teacher
American Association of Physics
    Teachers
One Physics Ellipse
College Park, MD 20740-3845

The Science Teacher
National Science Teachers Association
1840 Wilson Blvd.
Arlington, VA 22201-3000

STUDENT PROGRESS CHART

WORKSHEETS

Name	1	2	3	4	5	5a	T.A.	6	7	8	9	10	11-1	11-2	12	13	14
1.																	
2.																	
3.																	
4.																	
5.																	
6.																	
7.																	
8.																	
9.																	
10.																	
11.																	
12.																	
13.																	
14.																	
15.																	
16.																	
17.																	
18.																	
19.																	
20.																	
21.																	
22.																	
23.																	
24.																	